RAINY

by Robin Nelson

first step nonfiction

Lerner Publications Company · Minneapolis

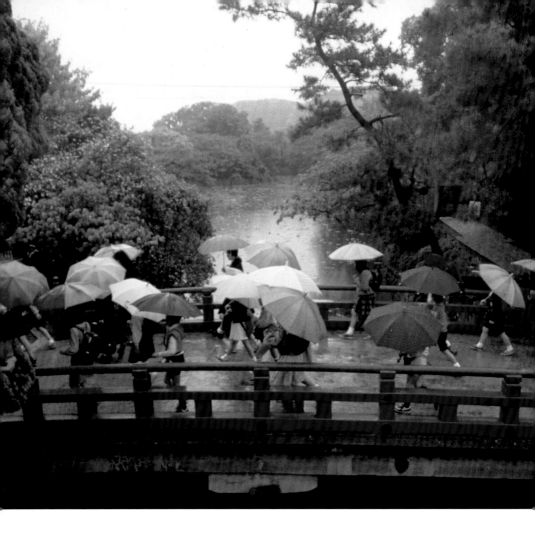

It is rainy.

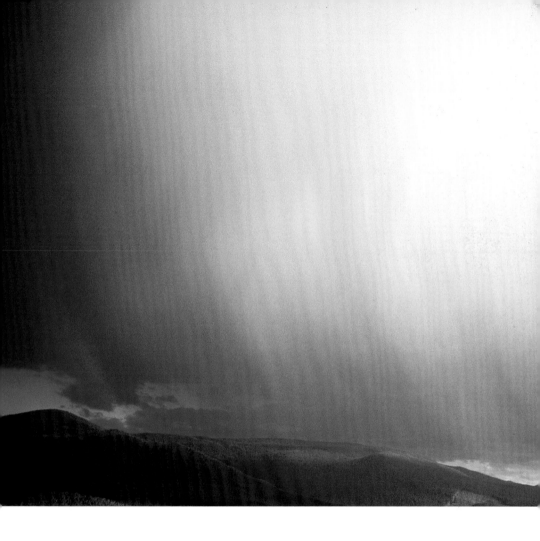

I see clouds.

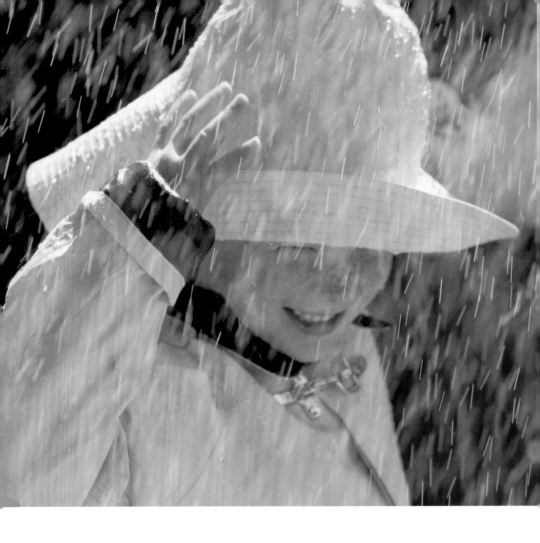

I see raindrops.

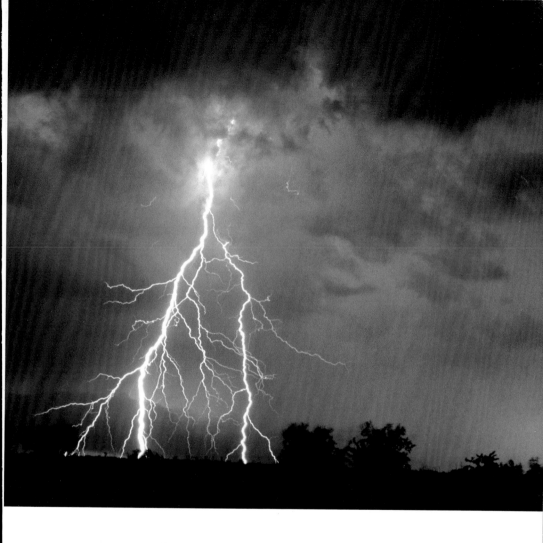

I see lightning.

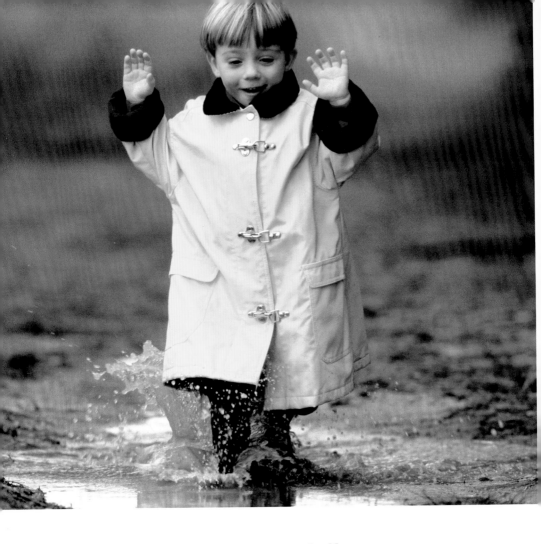

I see puddles.

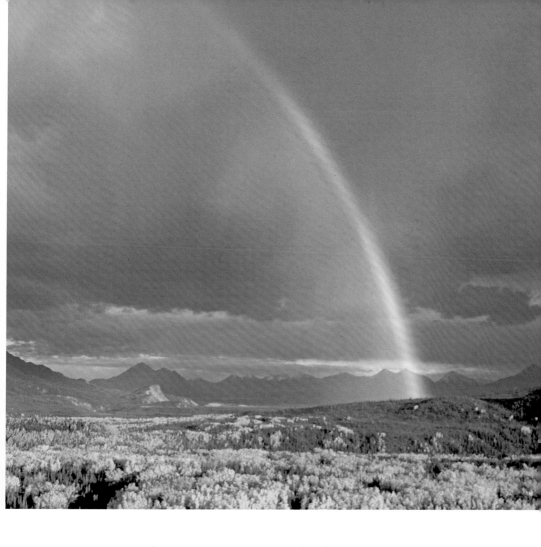

I see a rainbow.

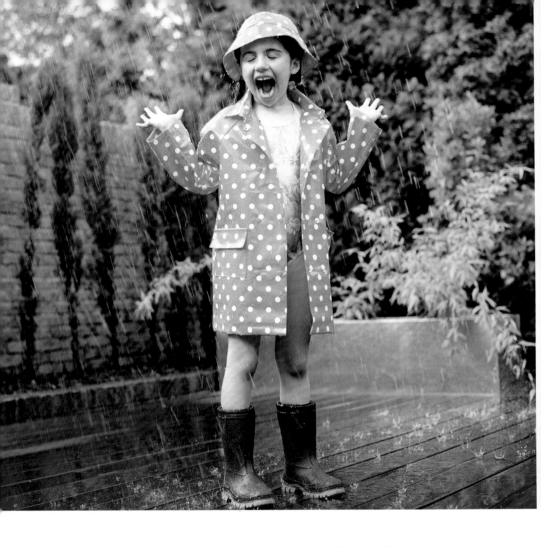

I like rainy days!